Growing Potatoes For Beginners

Beginners

Step by Step Blueprint on How to Grow Potatoes and Have A Thriving Potato Garden for An All-Year Supply

Introduction

Are you wary of store-bought potatoes? Learn How to Grow Your Own with Ease!

Are you tired of spending money on store-bought potatoes, often full of chemicals and pesticides?

Do you want to feel the satisfaction of growing organic potatoes?

Does the prospect of growing potatoes excite you, but you don't know where to start?

If you have answered YES, then:

Unlock the Secrets to Bumper Potato Crops with this Beginner's Guide!

Potatoes significantly augment the dietary needs of most people around the globe, and most of us consume them regularly, if not daily, in most households. They are versatile, nutritious, and filling.

With over 300+ available recipes online, you can enjoy them fried, roasted, baked, mashed, or in casseroles and stews.

They are widely commercially grown by farmers worldwide because they are easy to grow and in high demand.

In the United States alone, the average person consumes around 110 pounds of potatoes annually, with 30% consumed as French fries!

When we purchase our potatoes from grocery stores or supermarkets, we are usually limited to the common varieties, such as the Jersey potatoes, Russet potatoes, Maris Piper, charlotte potatoes, or red rooster. To access a wider variety of potatoes, you have to look for organic markets, which are not always accessible or close by.

Commercially grown and store-bought potatoes generally contain pesticide residues and other harmful chemicals. These residues may be carcinogenic and detrimental to human health.

For example, in a study conducted by the Environmental Working Group, potatoes had the third highest level of pesticide residue compared to other fruits and vegetables.

Further, a fertility study conducted by a team of researchers from Harvard University using USDA testing found that commercial potatoes have a high pesticide residue score. They also found that foods with high residue caused lower fertility and higher levels of urinary pesticides.

Growing and eating organic allows you to control the use of pesticides and chemicals, consequently ensuring sufficient nutrition and a lower pesticide residue-associated health complications risk.

So, why not grow your own?

Of course, this may seem easier said than done, but if you are wondering:

- ✓ *What is the best way to plant potatoes?*

- ✓ *How long does it take to grow potatoes?*

- ✓ *What is the best time to plant potatoes?*

- ✓ *How much sun and manure/fertilizer do potatoes require?*

- ✓ *How deep should you plant your potatoes?*

You have this easy-to-follow guide compiled with all the required information.

Expect to learn the following:

- Different types of potatoes.

- How to select seed potatoes.

- The best climate, soil, and seasons for potato farming.

- How to prepare your seed potatoes for farming: everything from green sprouting and cutting.

- Different potato growing methods.

- How to water your potato plants.

- Common pests and diseases affecting potatoes.

- Harvesting and storing potatoes.

- How to save your potato plants for a second harvest.

- **And much more!**

So, whether you want to grow various or specific potatoes, this detailed guide will help you learn everything you need to know about growing different varieties of potatoes!

Let us begin!

PS: I'd like your feedback. If you are happy with this book, please leave a review on Amazon.

Please leave a review for this book on Amazon by visiting the page below:

https://amzn.to/2VMR5qr

Table of Contents

Introduction 2

Chapter 1: 36 Different Types of Potatoes ...11

Starchy Potatoes 11

Waxy Potatoes 22

All-Purpose Potatoes 40

How To Decide Which Variety Best Suits Your
Needs 51

Chapter 2: How to Select Seed Potatoes 54

What to Consider Before Selecting Seed Potatoes
.... 55

The Best Places to Purchase Seed Potatoes 57

**Chapter 3: Climate, Seasons, And Soil for
Growing Potatoes 61**

Soil 61

Climate...64

Potato Planting Seasons........................64

**Chapter 4: How to Prepare Your Seed
Potatoes for Planting.......................... 67**

Chitting Your Potatoes......................... 67

Cutting Seed Potatoes 69

Protection From Frost 71

Chapter 5: How to Plant Your Potatoes 73

1: Planting Them Directly in The Ground- Without
Trenches... 74

2: Planting Them Directly in The Ground- With
Trenches... 76

3: Planting In Raised Beds.................... 78

4: Planting In Containers.....................80

5: Greenhouse Planting........................83

Potato Development Stages.................................84

Earthing Up Your Potatoes 87

How To Water Your Potato Plants 89

How to Harvest Potatoes..................................90

How to Store Potatoes...................................... 91

To Wash Or Not Wash Potatoes? 93

Replanting Potatoes The Right Way 94

Chapter 6: Common Pests and Diseases Affecting Potatoes...................................... 97

Bacterial Diseases..98

Fungal Diseases... 103

Other Diseases...113

Pests 120

Chapter 7: Common Potato Growing Mistakes to Avoid **135**

Conclusion **137**

Chapter 1: 36 Different Types of Potatoes

Before we get to the potato growing and care part, let us begin by understanding the available varieties because this knowledge will influence many other things in your journey.

Most of us have seen or consumed potatoes throughout our lives but have yet to consume most of the varieties. Potatoes have wide varieties and range in color, texture, taste, and shape.

Here is a guide to different types of potatoes, from the most common to those you will not find in your local grocery stores.

We can classify potatoes into three primary categories: starchy, waxy, or all-purpose.

Starchy Potatoes

These have high amounts of starch and break down quickly when cooked, making them ideal for mashed or baked potatoes.

1: *Russet*

The russet is the most popular potato variety grown in the United States, which is no surprise as its popularity comes with its many uses. Most restaurants and fast-food joints prefer the Russet for French fries due to its long oblong shape and for baked and mashed potatoes.

You can find the Russet potato all year round in your local grocery, although they are usually in season from mid-August to September.

There are two main types of Russet potatoes;

- Russet Burbank

- And Russet Norkotah

The two are very similar (you would not tell the difference by looking at them). They have minute differences.

Russet Burbanks have more flesh and store less water than the latter.

Russet Norkotah, on the other hand, has less thick flesh. Thus, many people prefer them for baking, roasting, and mashing.

2: *Yukon gold*

Yukon gold potatoes vary from medium to large size and are easy to recognize by their rough skin and golden-yellow flesh that has a buttery flavor when cooked. It is also all-purpose; you can mash, boil, roast, sauté, or fry it.

It may please you to learn that they are the most preferred choice for high-profile events such as White House dinner and the Oscars!

The Yukon gold potato is in season from August through February but is also available all year round with proper storage.

3: Kennebec

The Kennebec potato is common in the United States and is very popular because it produces high yields.

With its thin, smooth skin, and white flesh, the potato also withstands longer cooking and boiling time while at the same time maintaining its shape and enough starch for mashing.

You can use Kennebec potatoes for mashing, baking, or roasting, although they taste best when fried.

They are usually in season in the late summer through early winter, although they are available all year round when stored properly.

4: Yukon Gold Potatoes

The Yukon gold potato is the brainchild of Gary Johnson, who developed it in Ontario, Canada, in 1960. It is a crossbreed between the North American white and the South American potato.

Many consider this an all-purpose potato because it has a good texture between the waxy red potato and starchy russet potato. These features make this potato perfect for sauteing, frying, boiling, roasting, and mashing.

Identifying this potato is easy. Look for a potato with rosy pink eyes. These eyes distinguish it from other potatoes that are thin-skinned and have yellow flesh.

5: *Norchip Potatoes*

This potato is oval or round and has smooth, white flesh and light-buff skin. It was developed in 1961 by Robert H. Johansen from two potatoes – M5009-2 and ND4731-1.

This potato plant may mature earlier than others, its leaves are medium in size, and it is tolerant to diseases such as late blight, spindle tuber, leaf roll, verticillium wilt, and virus X.

6: Gold Rush Potato

This potato is a breed of two potatoes – a Lehmi russet and an ND450-3Russ.

It is a medium-sized russet cultivar; the plant has pubescent leaves, the tuber has white flesh, and the stems are green with some pigmentation at the base.

7: Hannah Sweet Potato

This type of potato is medium to large and cylindrical with rounded ends. This potato variety is slightly sweet and has a dense starchy texture consistent with a white potato.

Those who plant this potato do so because of its texture and flavor. It is also perfect for almost all types of dishes since it does not overpower the dish.

8: Japanese Sweet Potato

Known as Satsumaimo, Japanese sweet potatoes are starchy root vegetables with red-purple skin and pale cream flesh that turns yellow after cooking and are slender or smaller in shape. They originate from South and Central America, not Japan, as the name implies. However, they find extensive use in Japanese cuisines.

9: *Jewel Yam Potato*

The Jewel yam potato is an orange-fleshed sweet potato that dates to the mid-20th century in the United States. It has a bright orange moist, firm flesh and is cylindrical with copper-colored skin and tapered ends.

You can boil, bake, and mash this potato. However, it is essential to note that it gets watery when cooked since it is moist.

10: Norkotah Potato

North Dakota St. University released this potato variety in 1987. It is a medium potato that matures between 95 and 110 days from planting, has small roots, making it drought intolerant, and the tubers tend to be oversize.

This potato is severely susceptible to diseases such as silver scurf.

Waxy Potatoes

Waxy potatoes do not break down at all when cooked. Therefore, they are great for soups, roasting, stews, and salad.

They include:

11: French fingerling

French fingerling potatoes range from small to medium sizes and have rose-colored skin, waxy-yellow flesh with bits of pink, and an oblong shape.

Because of their delicate and edible skin, you do not need to peel them before cooking, making them ideal for baking, roasting, and roasting with their skin on.

They are usually in peak season in late spring through early summer but are also available year-round.

12: New potatoes

Like their name, new potatoes are young because they are harvested early in the season. The mature ones, however, are harvested when their tops turn brown. This variety is special because it has very thin skins and very moist flesh, a rare combination in potatoes.

You can enjoy these potatoes during spring and summer, but they do not store well due to their high moisture levels. Therefore, you should consume them within two weeks of harvesting.

13: Russian banana potato

This is a fingerling variety (varieties with small, stubby, long potatoes) that includes the French fingerling and purple Peruvian known for their small and narrow size that derives its name from oblong, crescent shape, and yellow shape.

Like other fingerling potatoes, it has a delicate thin skin that does not require peeling before cooking and can be ideal for cooking methods such as roasting, grilling, baking, steaming, and boiling. Its unique shape and sweet flavor make it a preferred culinary choice but unsuitable for mashing or frying.

The Russian banana fingerling potatoes are usually in season in the late summer and early fall but are available year-round. They also store well for 3 to 5 weeks in cool, dry, and dark places.

14: Charlotte potatoes

Also known as salad potatoes, these relatively small potatoes are popular for their salad use and are widely available in most grocery outlets. You can identify them by their creamy skin and light yellow or white flesh with a mild flavor.

One benefit of growing them in your garden is that they continue growing well even when left in the ground longer than necessary. The crop also grows well in most soils and has excellent disease resistance.

Charlottes are in season in the late winter through spring and late summer through early fall but are also available all year round.

Due to their mild flavor, they work best when incorporated into other meals or salads.

15: *All Blue potatoes*

As the name suggests, the All Blue variety is blue from the interior to the outside and is not as common as other blue varieties. Look for a pale ring in their blue flesh to identify them from other blue varieties. The crop also has a good reputation for being exceptionally drought-tolerant, although it may not do well in wet seasons.

The All Blue potatoes are all-purpose, making them suitable for various cooking methods, but if you want to preserve their color, you may as well steam or bake them.

Other names of the All-blue potato include Fenton Blue and Vitelotte.

16: Purple Peruvian

Purple Peruvians are very common in South America, where they originate, but rare in other countries. Like the name, the potato has purple skin and flesh and a small, slender shape. Its flesh is also dry and starchy, making the variety ideal for mashing or deep frying.

17: Red Bliss Potato

This potato has a thin, smooth red skin and white flesh. It has a low starch content, is creamy, and the flesh is moist. Those who use it prefer it because it holds its shape perfectly after cooking; it is also perfect for boiling, roasting, casseroles, and in potato salads.

18: Russian Banana Potato

This type of potato is 3 to 4 inches long and has thin skins with firm and buttercream-dense yellow-colored flesh that is quite delicious. Those who have eaten this potato variety prefer it because it keeps its shape well, slices neatly, and cooks up solid.

19: Red Adirondack Potato

Cornell University developed this potato. It has slightly flattened tubers, lightly netted purplish-red skin, and pink-red, moist, and flavorful flesh.

20: Creamers

This type of potato can be a red potato, Yukon, or any white potato, but it is known as a creamer because it is harvested young and before it grows into its full size. It is small, slightly oval, round, and tender.

21: Charlotte Potato

This potato is long and considered a salad potato. It is large and smooth with shallow eyes, thin light brown skin, slightly waxy, and yellow cream flesh. Those who love it prefer it because it's easy to grow and does so in most soils. Also, it is disease resistant, holds its shape well when cooked, and is delicious cold or hot.

22: Maris Peer Jersey Royal Potato

This small potato has a long, oval to kidney shape, a smooth rounded end, pale yellow skin with some speckling, cream-colored flesh, and is low on starch and high moisture. After cooking, the peer jersey royal potato is tender and has a sweet, earthy, and subtle flavor.

23: Bintje Potato

This potato originated in the Netherlands and was released in 1910. It is small to medium, has golden skin and yellow flesh with a silk-like finish, and is smooth and well-rounded. It is virus resistant but can get diseases such as late blight and scab.

24: Asterix Potato

This potato breed was developed in Holland by a breeder called HZPC. The plant has dark green leaves and produces many tubers (10 to 12); the skin is red, has yellow flesh, is thick, cylindrical, and has a terminal bud weighing between 70 and 120 grams.

This potato is susceptible to diseases such as tuber phytophthora, leaf enrolment, Y virus, and A virus but is resistant to warty scab and X virus.

25: Viktoria Potato

This potato is mainly firm to floury.

The plant is considered good to very high yield, the tuber is large and long, and its skin is yellow. It is sensitive to diseases such as scabs, viruses, and potato cancer, and its primary use is cooking French fries.

26: Maria Potato

This potato is perfectly round, has white flesh, is lightly textured, holds color exceptionally in storage, and has shallow eyes. It is resistant to diseases such as common scabs and golden nematodes, but, on the other hand, it is leafhopper and tarnished plant bug resistant.

27: Annabelle Potato

This potato is small to medium and has a long to oval appearance. The tuber has thin, smooth skin with shallow flat eyes, pale yellow or yellow-tan skin, and golden yellow flesh that is trim, dense, fine-grained, and slippery.

All-Purpose Potatoes

These potatoes are popular because of their ability to hold their shape when cooking, and as the name suggests, they are all purposes.

28: Red gold

This variety has delicate thin, rose-red skin and yellow flesh; like the new potato, it produces a high yield. It is versatile because you can mash, roast, fry, or steam it.

The red gold potatoes are in season during summer through spring but can grow all year round under moderate climates.

29: Elba

This yellow-colored variety has flaky skin and moist flesh, making it great for mashing, boiling, and baking. One of its great qualities, and for that matter, benefit, is its resistance to viruses, early and late blight, and scab diseases.

The crop is also drought-tolerant, performs well during dry times, and stores well.

Since it is an all-year-round crop, you can enjoy it any time of the year.

30: Jewel yams

The name yam has nothing to do with the potato being a true yam; instead, it is a nickname for a sweet potato variety.

Jewel yams range from medium to large sizes and have a cylindrical shape with tapered ends and rough, brown-rose skin with dark spots with faint lines. The exterior has a covering of tiny root hairs, and the flesh is orange, moist, and firm.

The peak season of jewel yams is fall through winter, but the variety is also available year-round.

Due to its starchy, moist, and fluffy texture, you can use the potatoes for cooking boiled, roasted, steamed, and mashed potatoes.

31: Red Pontiac

With their red skin, and sweet, white flesh, Red Pontiac potatoes are perfect for growing as new potatoes (they do not have to wait until fully matured for harvesting). Unlike most potatoes that grow well in loose, well-drained soils, this variety performs well in heavy soils like clay, making it a favorite among gardeners.

The peak season for Red Pontiac potatoes is in late summer, but since they store well, you can enjoy them during the cool months of winter and fall.

You can use them for mashing, making French fries, boiling, baking, roasting, and potato crisps.

32: German Butterball

German Butterballs range from medium to large sizes and have golden skin with yellow, tender, moist flesh and a round to oblong shape. The smooth skin is also lightly covered with dark brown spots, brown patches, and shallow eyes.

The potato is usually in season in late spring through late summer and is available year-round.

Use German butterballs for cooking applications such as baking, steaming, frying, roasting, and mashing.

33: White potatoes

This variety ranges from small to medium with a round to long shape and white and tan skin and flesh. These potatoes also have thin, delicate skins that give mashed or roasted potatoes a nice texture without peeling.

34: *Norland Red Potato*

North Dakota State University developed this potato. It produces medium to large tubers; the skin is red and has shallow eyes, the flesh is light-yellow to white, and it is resistant to scabs.

35: Purple Majesty Potato

Developed in the Colorado mountain area., this potato is dark purple (flesh and the skin), has a height of 18 to 24 inches, and is perfect for boiling, baking, and frying since it has a low starch content.

36: Peruvian potato

Initially found in the Andean highlands of Peru, Chile, and Bolivia and dating back 13,000 years, these potatoes come in different forms —there are over 4,000 types, colors, and flavors.

All these varieties have one thing in common: they are nutrient-dense and a good source of fiber, potassium, magnesium, and carbohydrates.

Peruvian potato variations include, but are not limited to:

- Papa blanca

- Papa amarilla

- Papa huamantanga

- Papa purpura

- Papa peruanita

- Papa tarmena

- Papa coctel

- Papa rosada

- Papa perricholi

- Papa huayro

- Chuno

- Oca

- Camote

The list of potatoes you can plant in your garden is endless, but you must decide what kind of potatoes you will grow in a particular year. You can plan that around December or January after preparing your garden for planting.

Now that you have a working knowledge of the diverse varieties of potatoes, it's easier to identify the ones you will grow.

How To Decide Which Variety Best Suits Your Needs

Here are some key things to consider when deciding which potato variety to grow in your garden:

1: Intended use

How do you usually prepare your potatoes?

Give some thought to how you usually prepare your potatoes, then make your choice regarding the three categories:

- Starchy

- Waxy

- All-purpose.

Note that the amount of starch differs in different varieties of potatoes; while some break down into fluffy textures, others remain firm and waxy, which also determines their uses.

Starchy potatoes are dry, making them easier to break down than waxy potatoes, which is why they are ideal for mashed and baked potatoes. Additionally, they have an external layer that absorbs oil and turns crispy while the inside remains fluffy, making them great for fried potatoes.

Waxy potatoes are the complete opposite. They have a high-water content, so they do not absorb liquid like starchy potatoes and do not break down when boiled or cooked in liquid. This fact makes them more suitable for soups, salads, and stews.

Finally, all-purpose potatoes have medium moisture and starch content making them perfect for any waxy or starchy potato recipe. Also, when baked, these potatoes are creamy, and when boiled, they hold their texture.

2: Location

You can ask other potato gardeners or sellers in your region about the potato varieties that do well in your area. Knowing which types do well will help you choose a type not commonly affected by diseases and pests.

Cases of certain potato diseases are common in some areas. You don't want to grow a variety that may be problematic in your area, which means more maintenance work and usually lower yields.

Also, by knowing the common diseases affecting potatoes in your area, you can find potato varieties resistant to specific diseases, early maturing, and stress tolerant.

3: The length of time before harvest

Different potato varieties mature at different times. In terms of time before harvest, there is the *first early variety*, followed by the *second early variety*, and finally, *maincrop potatoes*. We will discuss this in more depth in subsequent chapters

Let's now focus on where to get potato seeds:

Chapter 2: How to Select Seed Potatoes

As a potato grower, one crucial decision you make when getting started on potato growing is selecting good potato seeds, an effort you must make to have a healthy crop.

But first, let's explain what seed potatoes are:

Seed potatoes are tubers grown for replanting to produce potatoes. Most potato growers save them from the previous season's harvest and store them under certain conditions to prevent diseases and keep them firm.

Never plant storage-sprouted potatoes because they may have disease and fungus. It is important to note that a diseased seed potato will do poorly and transfer the disease to subsequent plants by contaminating the soil. Therefore, never leave your leftover potatoes to sprout in your garden.

So, how can you make good decisions when selecting seed potatoes? The answer is simple:

What to Consider Before Selecting Seed Potatoes

As you consider what variety of potatoes to grow, think of the following:

Buy disease-free seed potatoes

As a gardener, you should only purchase seed potatoes certified as disease-free rather than buying potatoes from grocery stores or using potatoes from the garden. That way, you will not have to worry about diseases like blight destroying your entire crop or using a few pesticides to prevent it.

You may opt to save on costs by deciding to transplant potatoes from a previous harvest. Still, as we have already seen, some potatoes may carry diseases from the last season and may spread diseases to the next crop.

Purchasing organic seed potatoes

Purchasing organic seed potatoes is the other option. You can buy them from several places, including farmer's markets, garden centers, and online retailers. To ensure you buy the best organic potatoes, here are some tips for you:

- Buy certified organic seed potatoes grown using organic methods.

- Ask the seller how to care for and store your seed potatoes to ensure they sprout and grow well.

- Buy locally to avoid damage from impact during transportation.

However, it would be best to consider a few things before going organic.

- First, organic seed potatoes cost more than non-organic ones, so consider your budget.

- Second, organic seed potatoes are not as disease-resistant as non-organic seeds, and finally, organically seed potatoes are not readily available.

That leaves us with the question: where to get excellent and reliable potato seeds?

The Best Places to Purchase Seed Potatoes

Here is a guide to the best places you can purchase your seed potatoes.

Locally

Buying your seed potatoes locally will depend mainly on where you live. In regions with only small seed potato retailers, factors such as the unavailability of good seeds in stores during planting time or when required may cause farmers to plant what is readily available.

Also, in other regions, you may find that the good seed potatoes available in stores are limited to varieties that thrive in the given areas; therefore, one may have to look elsewhere for other types. That is unless you have larger seed potato retailers in your area.

However, suppose you still opt to purchase your seeds locally. In that case, the farmer's market and seed exchanges (where gardeners meet and exchange their seeds) are reliable places for various seed potatoes. However, schedule ahead for the trip. During these events, the chances of getting rare varieties from other potato farmers are greater than from local stores.

Note that seed potatoes from farmer's markets are mostly organic.

Online

If you cannot find anyone selling potato seeds in your locality, you should consider buying them online. This is convenient because the seller can ship them to your address. All you need to do is to know which variety to go for when placing an order:

Here are some ideas on where to get your potatoes online:

Amazon[1]

Gurney's seed and Nursery co[2]

The Maine potato lady[3]

Wood Prairie Family farm[4]

---

[1] https://www.amazon.com/seed-potatoes/s?k=seed+potatoes
[2] https://www.gurneys.com/category/potato-plants-and-sets
[3] https://www.mainepotatolady.com/productcart/pc/viewPages.asp
[4] https://www.woodprairie.com/awesome-two/

Johnny selected seeds[5]

Urban farmer[6]

Seed exchanges can also happen online, or people can email each other, especially because some participants do not live in the same region. Potato farmers simply arrange a seed swap event online, and then the gardeners offer to trade whatever seeds they have for whatever seeds they want.

Some of the online seed exchanges for potatoes include;

Seed savers exchange[7]

Southern exposure seed exchange[8]

As you purchase your potatoes online, you should consider these things:

- Check the details on each variety to know things like harvesting time to ensure you do not buy overstayed seed potatoes.

---

[5] https://www.johnnyseeds.com/vegetables/potatoes/
[6] https://www.ufseeds.com/vegetables/potatoes
[7] https://www.southernexposure.com/categories/sweet-potatoes/
[8] https://www.southernexposure.com/categories/sweet-potatoes/

- Ask how long they have been in storage because storage should not exceed 5-10 months, depending on the storage method.

- Ask other gardeners who have bought their seed potatoes online for recommendations.

- Only buy seed potatoes certified as disease-free to avoid planting disease-contaminated seeds.

- Ask about delivery to your region.

Next, let's look at the most favorable soils, climates, and seasons for growing potatoes.

Chapter 3: Climate, Seasons, And Soil for Growing Potatoes

This chapter discusses the right potato-growing climate and soil and what to do if you do not live in places considered favorable to growing potatoes.

Soil

Potatoes require loose sandy soil that is very fertile with a PH of around 5 to 6. The soil should also be well-draining but also moisture retaining. The moisture supply must also be consistent throughout the growing season so the tubers form to size.

NOTE: Planting potatoes in poorly drained soils like clay or silt tends to result in lumps.

Potatoes with lumps

Also, if the PH is above 6.0 or below 5.0, your potatoes will not thrive because they will not absorb enough nutrients, and their immunity will reduce, making them prone to a disease called potato scab.

However, if your soil does not meet the appropriate conditions ideal for growing potatoes, it does not mean you cannot grow potatoes.

Clay soils, for example, do not meet the potato growing conditions above, but there are ways you can improve them. One way to do so is by adding aged manure or compost to help loosen up the soil and give it better drainage.

Manure or compost also adds some organic material and nutrients to the soil, making it fertile. As we have seen, sand soil is ideal for potato growing, but you should not consider adding it to clay soil as it will make it very hard and challenging to work with.

In cases where you may have loam soil in your garden but it is not as loose as recommended, you can add some sand soil to it to loosen it more.

If you need to correct soil PH, there are also ways to improve it depending on whether the PH is more acidic or alkaline.

To add acidity to your soil, you must lower the PH by amending it with agricultural sulfur, which you can find in most of your local garden supply stores. Depending on the soil type, the seller will advise you on how much sulfur you need to add to your garden soil. After that, you will have to wait at least a month before you plant your potatoes because concentrated sulfur can cause your potatoes to have a stunted growth or dark color.

On the other hand, if your soil is too acidic, you will need to make it more alkaline by adding some agricultural lime or wood ash. However, use wood ash because it is readily available and contains nutrients such as calcium and potassium that can benefit your crop growth. Note that you should start with small quantities of any soil additives and test the PH before and after adding using a litmus paper or PH-testing kit.

Climate

Although potatoes can grow in most climates, they do well in areas with ample rainfall, cool weather, and a temperate climate.

Temperatures above 80 degrees Fahrenheit are usually too warm for planting and growing potatoes.

Potato Planting Seasons

The key things you need to know here are:

Spring planting

If you live in places that experience winter, the best time to plant your potatoes is in early spring, around two or three weeks before the frost clears.

Summer planting

You will harvest these potatoes during autumn, so plant them about 12 weeks before the first frost.

Winter planting

If you live in places where the winters are mild, but the summers are hot, plant maincrop potatoes in winter to harvest in mid to late spring or early season potatoes in late summer for harvest during fall. If you grow your potatoes in winter, they must get enough light and protection from hard frosts and freezes.

In regions with harsher winters like Dakota, Michigan, Alaska, and Minnesota, you must avoid planting your potatoes outdoors when winter comes. However, you can still grow your potatoes in containers indoors or in heated greenhouses because heavy snow or hard freezes can damage your crop.

Tropical and sub-tropical planting

On the other hand, if you live in tropical and subtropical regions, the best time to do so is in late summer, spring, and autumn for harvest before the rains start. This is because the humidity level in early summer hinders the successful growth of potatoes.

If the climates are challenging to comprehend, try viewing it this way;

- First, we plant early varieties between February and late May and harvest them ten weeks after planting. Examples of first-early varieties include Red Norland, Rio Grande Russet, Red gold, and Irish Cobbler.

- We plant second early potatoes between early March and April and harvest them shortly after the first early potatoes, around June and July. Examples of this variety are Yukon gold and red Pontiac.

- On the other hand, we plant maincrop potatoes in mid to late April because they need more time to mature — around 15-20 weeks after planting. Examples of maincrop varieties include Russets, Elba, Kennebec, French fingerling, German Butterball, charlottes, All-Blue, purple Peruvians, and Russian fingerling.

Let's look at how to prepare your seed potatoes for planting.

Chapter 4: How to Prepare Your Seed Potatoes for Planting

First, always prepare your seed potatoes before planting so the roots develop properly, and the plant takes off. After deciding which potatoes you will grow and ordering your seed potatoes, you can prepare them for planting around Mid-January.

Below are the preparation methods necessary before planting:

Chitting Your Potatoes

Also known as green or pre-sprouting, chitting is a potato preparation method that encourages seed potatoes to sprout before putting them in the ground. The method gives your tubers a head start and triggers faster growth and a higher yield once you plant them.

Although you can plant your seed potatoes in the soil exactly as they are from the store, chitting them means you will enjoy an earlier harvest or, better yet, a second crop. The seed potatoes will require 4 to 6 weeks of chitting before they are ready to plant, and you can start doing it in January.

1. Take your seed potato and identify the most sprouted side. This is the end where your potatoes will sprout from, which is also known as the **rose end**. Count at least four eyes or sprouts, then remove any other sprouts so your potatoes grow in one direction.

2. After removing the excess sprouts, arrange them in seed trays or egg trays with the rose end facing up and the heel (the narrow end of the potato where it is cut from the vine) sitting in the box.

Seed potatoes in trays

3. Label your potato seeds according to the variety and date you bought them so you do not confuse them; then look for a bright, dry room with cool to warm temperatures to place them. Doing this encourages strong green shots instead of weak white ones. Weak white stems result from placing seed potatoes in a dark room. You can leave them on a window sill or place them outside if the room is not bright enough.

4. Continue to check on your seed potatoes to ensure the shoots are healthy as they develop and you are not exposing them to excess light, heat, or moisture. Also, check if any potato has mold and remove it to avoid affecting the healthy ones.

Cutting Seed Potatoes

Although you will mostly buy seed potatoes the size of chicken eggs, you will sometimes get a seed potato that seems too large to plant. The trick is to cut it into two or three pieces.

As you cut your large seed potato into manageable pieces, you must ensure that each piece has several sprouts because these develop into shoots; at least three sprouts per piece will do.

Here is how to go about it:

1. First, chit them as we learned earlier, and then two days before you plant them in the soil, cut them. As a rule, use a clean, sharp knife to cut your seed potatoes.

2. After cutting the large seed potatoes, please leave them in a cool and dry place for 24 to 48 hours, with the cut edges facing upwards so they dry and create a protective membrane. They are likely to rot if you do not leave them to dry.

Cut seed potatoes

If you plan to postpone the planting, perhaps due to weather changes, ensure you place your seed potatoes in a cooler location to slow down the growth of the sprouts.

Protection From Frost

Seed potatoes are prone to frost damage as they are usually tender. As such, always ensure you protect them from frost. Frost will damage the shoots and kill the entire seed potato. As such, whether you chit them or not, keep them indoors in a space that meets the conditions suitable for their storage if there is a risk of frost.

Keep monitoring your chitting seed potatoes through January and February, and reject tubers showing signs of diseases.

As soon as the sprouts are ½ inch to 1 inch, they are ready for planting. Let us see how in the next chapter.

Chapter 5: How to Plant Your Potatoes

You can start planting your sprouted seed potatoes in late March or early April, so ensure everything is ready, including your tools, manure, and compost.

There are several ways you can grow potatoes depending on various factors, including the space available, how many potatoes you want to grow, and several other factors. They include:

- Directly on the ground – without any trenches.

- Directly on the ground – with trenches.

- In raised beds.

- In containers.

- Greenhouse planting.

Let us learn more about these methods:

1: Planting Them Directly in The Ground- Without Trenches

Pros:

- ✓ You will not have to dig up potatoes.

- ✓ The potatoes grow clean and smooth compared to those in the soil.

- ✓ No weeds, as the mulch blocks out the light.

Cons:

- ✓ This method is unsuitable for early spring as the frost will damage the crop.

- ✓ The plants may need a bit more watering.

- ✓ Potatoes grown this way are prone to slugs.

How to do it:

1. First, dig the area you want to grow your potatoes using a hoe to clear the weeds and loosen the soil. Add a layer of compost on top.

2. Next, water the entire area well, then plant your sprouted potatoes by spacing them on the surface. The spacing should be 12 inches in every direction.

3. After placing all your sprouted potatoes in place, cover them with 2-6 inches of straw or a couple of inches of other compost.

4. Water the straw thoroughly because potatoes grown this way tend to dry faster.

5. You can keep the wind from carrying away the straw by covering it with mesh.

6. Once the plants start to grow past the straw, remove the mesh and check if the straw is holding some plants back so you can help them through it.

2: Planting Them Directly in The Ground-With Trenches

With a well-prepared garden, you can now dig the trenches before planting.

Pros:

   ✓ Potatoes get more room to develop because you bury them deep in the soil.

Cons:

- ✓ During rainy seasons, the trenches fill up with water, which may cause the tubers to rot.

- ✓ The trenches tend to fall on top of new plants, which can smother them.

Here are simple steps to dig proper trenches suitable for potato planting and growing:

1. First, dig some shallow straight trenches about two to three feet apart. The trench should also be about 4 to 6 inches deep so the potatoes have enough room for growth.

2. Plant your potatoes 12 inches apart and cover them with 3 inches of soil.

3. When your shoots reach about 12 inches long, scoop soil from between the rows using a shovel and pile it against the plants to bury the stems halfway. Keep doing that throughout the growing season.

3: Planting In Raised Beds

Due to the spacing of the beds, the ideal potatoes to plant in raised beds are the French fingerling because they are small; therefore, they will not take up more space when they begin to grow.

Pros:

- ✓ Potatoes have a higher yield because the soil is looser.

- ✓ The soil warms up faster and stays warmer longer, lengthening the growing season in fall.

- ✓ You can control the soil content in the beds.

✓ This is an excellent growing method in areas where the soil is compact and poorly drained.

✓ Fewer weeds.

Cons:

✓ The dense spacing in the beds can restrict nutrition.

✓ It takes much work to fill the beds up with soil.

To plant your potatoes in raised beds;

1. Start by filling your raised beds halfway with fertile soil. Mixing the soil with some compost or manure to add nutrients to the soil and loosen it is a good idea. Also, ensure your beds are in an area that receives adequate sunlight for about 6-8 hours daily. The beds should be at least 12 inches deep.

2. Next, space your sprouted potatoes about 12 inches in all directions and then bury them 3 inches deep by covering them with three inches of soil.

3. Once the potatoes grow, add more soil into your beds until they are full.

Alternatively,

4. Make trenches about six inches deep in the soil, and space them about 12 inches apart.

5. Place your sprouted seed potatoes in the trenches, ensuring they face up and spaced 12 inches apart.

6. Add compost or manure on top, then cover the seed potatoes loosely with three inches of soil.

4: Planting In Containers

This is the best way to grow potatoes in limited spaces. You can even place the containers on your balcony or patio, which will still do well.

The ideal potatoes to grow in containers are the smaller varieties, like the fingerling and red potatoes; the larger varieties, like the Russets, will need more space to expand to full size in a container.

Pros:

- ✓ It is easier to monitor them for critters.

- ✓ They produce a good yield and grow fast.

- ✓ Harvesting is easy.

- ✓ You can bring your containers indoors if there is a later spring frost.

Cons:

- ✓ They need regular watering.

- ✓ They have a lower yield.

- ✓ The sprouts do not produce large tubers.

Get a large plastic bucket or grow bag 2 to 3 feet tall with a 10-15 gallons capacity. Other container options include garbage bin bags, canvas tote bags, wooden half-barrels, and sacks if they have drainage holes at the bottom.

Nowadays, special potato pots with harvest doors have become more popular as more gardeners opt to purchase them. However, ensure you pick a quality pot with doors that seal tightly to prevent soil from drying out too fast.

Also, get high-quality potting or organic soil from your local garden center. If you prefer using your garden soil, add nutrients and manure to loosen it.

Look for a spot that gets at least six hours of sunlight daily, place your container/s there, and then fill it with about 6 inches of quality soil.

Next, place your sprouted seed potatoes with the eyes facing up on the soil surface. Space them at least 5 inches apart in all directions. The width of your container determines how many seed potatoes you plant regarding spacing

Once done spacing your seed potatoes, cover them with 4 to 6 inches of quality soil.

Keep adding soil around the plant as it grows to increase your yield. Then, once your plant reaches 8 inches tall, add 2 inches of quality soil to your container/s.

5: Greenhouse Planting

You can also grow your potatoes in a greenhouse, especially in an area prone to heavy rain.

Pros:

- ✓ You can grow potatoes out of season.

- ✓ You can grow your potatoes in winter.

Cons:

- ✓ Limited amount of space, considering potatoes need more growing space.

- ✓ They need watering much more often.

- ✓ The humid environment in a greenhouse can cause fungal problems.

If you prefer to grow your potatoes in a greenhouse, you should avoid planting them directly in the soil because they will be hard to dig up without interfering with your greenhouse. Instead, plant them in containers.

Since you will be growing your potatoes in containers, the method for planting is the same as container planting, with the only difference being that the containers will be in a greenhouse.

Let's now focus on a potato's developmental stages:

Potato Development Stages

Now that your potatoes are in the ground, containers, or beds, let us determine what stages they pass through before harvest time.

There are four main development stages known[9];

1. The tillering stage or vegetative stage.

2. The tuber formation stage.

---

[9] https://bit.ly/3z03e7H

3. The tuber bulking stage.

4. The tuber maturation stage.

Let us discuss each one of them in depth:

The tillering or vegetative stage

This first stage is all about producing robust large plants. The plants first produce roots, stems, and leaves (also known as vegetative growth) and remain in this state for 30 to 70 days.

Warm and moist weathers favor the vegetative development of potatoes (the warmth goes as high as 27 degrees Celsius or 80 degrees Fahrenheit and the moisture measures between 60 to 70 degrees Fahrenheit or 15.5 to 21 degrees Celsius).

You can these temperature combinations naturally if you plant your potatoes during spring, or you can induce them by adding organic mulches to the soil. Doing the latter will cool the soil, especially if you plant in early summer.

The tuber formation stage

The second development stage is tuber formation which takes around two weeks (the tubers grow in two weeks), and branching of the stems, which follows.

We cannot conclusively say how many tubers you should expect because this growth depends on the temperature, daylight hours, and water availability in the two weeks.

However, for optimal tuber growth, ensure the quantity of water is 5 gal/yd, and the temperatures at night should be 12 degrees Celsius or 54 degrees Fahrenheit.

The potato plants should also experience short day lengths (day lengths refers to the length of darkness that plants experience—short day refers to a length shorter than 12 hours). It is also essential to note that during this stage, flowering might happen, but this is not always the case.

The tuber bulking and maturation stage

During this stage, the tubers become bulky because the plant directs its energy to bullish growth while they are still in the soil.

This stage requires boosted nutrients and water (1 inch or 2.5 cm of water every week) intake, which can go as high as 90 days. The temperature should also be around 18 degrees Celsius or 65 degrees Fahrenheit. If the temperature goes above or under this temperature, the potato size will decrease by 4 percent.

Some farmers harvest their potatoes before they are fully mature. This is not ideal or recommended because the harvest will be short-term. In general, wait until your potatoes grow to full size, aka right after your potato tops yellow and die.

To be more specific:

- Potatoes planted between mid-March and mid-April should be ready for harvest after 10 to 12 weeks.

- Potatoes planted in April should be ready for harvest after 10 to 12 weeks.

- Maincrop potatoes planted in April will be ready for harvest after 15 to 20 weeks.

Earthing Up Your Potatoes

Earthing up is the practice of heaping soil above the planted potatoes to help increase the potato yield. Do this two or three times a season: once every two weeks after the potatoes have emerged full and twice after six weeks.

When growing potatoes, you will see them protrude from the soil as the tubers expand and push through the soil, which happens as the season progresses. If this happens, parts of the tubers will start to turn green and poisonous.

So, if you are wondering why earthing is necessary and how it helps increase potato yield, well, here is how:

- If appropriately done, you cover stems above the ground (stolons) with soil, thus providing necessary growing conditions that encourage these stems to produce tubers instead of leaves and, in turn, increase the number of potatoes produced.

- Earthing up allows air supply into the soil for sufficient root growth.

- The soil particles loosen, providing more room for the tubers to grow.

- You also get to remove any potential plants that are weeds.

However, be careful when earthing up to avoid damaging the roots or destroying the entire plant.

How To Water Your Potato Plants

Potatoes need plenty of water throughout the growing season as they are quite sensitive to lack of it. If you do not water your potatoes as required, the tubers will not develop as expected, and you will harvest tiny potatoes. Inconsistent watering can also cause hollow hearts, a potato disorder that occurs internally, causing a star-shaped cavity in the tuber.

To water your potatoes properly, give them a good soak of 2 to 3 inches of water once a week so the water reaches the tubers as they have deep roots. However, you do not have to water your potatoes if it rains a lot unless you grow them in containers or a greenhouse. Also, to prevent fungal infections, ensure you only water the roots rather than the leaves because getting water on the leaves will encourage infections.

Eventually, when the leaves dry up, you can slow down on the watering but do not stop entirely. Only water less frequently, but ensure the water is enough to keep your garden moist. Be careful, however, not to overwater your potatoes because too much water causes the tubers to rot and the leaves to wilt, thus damaging the whole plant. Stick your finger into the soil to prevent overwatering your potatoes. If it feels damp, then do not water your plants; wait a while and repeat the experiment.

Mulching your potatoes right after planting them is an effective way to conserve water as it prevents the water from evaporating, especially in hotter climates. Add about 3 inches of straw, leaves, or dry grass clippings to your crop; this way, you will not need to water a lot.

Learning how to water the earth and assess the potatoes' growth development is insufficient. We also need to know how to harvest them correctly when the time comes.

How to Harvest Potatoes

To harvest your potatoes, use a fork or shovel to dig up the soil about 18 inches from the edge of the potato leaves.

After removing one shovel, check for potatoes by feeling for their presence either by sight or hand. If you do not feel or see any potatoes after the first dig, dig a little further and closer to the plant, then assess the soil again.

Once harvested, the next thing to do is to store the potatoes.

How to Store Potatoes

There are several ways to store potatoes to ensure they last for months. But before you store them, you need to sort them.

Look for potatoes that do not have large blemishes or puncture marks (if you identify potatoes with such marks, eat them within a few days). Also, if you planted potato varieties with thick skin, such as russet and brown potatoes, note that they will store faster than red-skinned potatoes and fingerlings.

After storing them, choose one of the following ways of storing your potatoes.

Newsprint storing

Store them on a thick paper such as newsprint or in a space with a temperature of 44 degrees Fahrenheit or 6.67 degrees Celsius. In simpler terms, store them in a cool room with low humidity and light for around two weeks because storing them for this long thickens the potatoes' skin and closes off any small cuts.

Kindly maintain this temperature because if you store them at lower temperatures, they will lose their flavor and become starchier. On the other hand, if you store them at higher temperatures, the potatoes might spoil faster.

Refrigeration Storing

We advise against storing harvested potatoes in a refrigerator because it boosts their sugar levels, changing their flavor. Store them in the fridge if fully or partially cooked because it reduces browning enzymes, preventing discoloring. Also, if you identify injured potatoes, consume them first because they will not last long.

The following is the correct way of storing potatoes in the refrigerator;

- Peel your potato. Chop up large potatoes.

- Rinse the potatoes properly

- Put them in cold water, ensuring full submersion in the water (this will prevent them from turning brown)

- Bring a pot of water to a boil, then put the potatoes in. Let them boil for five minutes.

- Remove the potatoes from the boiling water, then put them in a container of ice water.

- Put them in freezer bags or vacuum sealers.

- When ready to eat them, pull the sealers or bags from the refrigerator, allow them to defrost overnight, then cook.

Do not store potatoes in light because it causes the potatoes to produce chlorophyll which turns the potato's skin color green. This green color might be harmless since it affects the potatoes' peel and the first 1/8th inch of the flesh, but excessive amounts produce a toxic chemical called solanine. Solanine makes the potatoes bitter and causes a burning sensation in the throat or the mouth, especially if you are extra sensitive to it.

To Wash Or Not Wash Potatoes?

Since we grow potatoes below the soil, do not wash them before storing them because water adds moisture which is the leading cause of bacteria and fungus overgrowth. Only rinse and scrub them when you are ready to use them. You can use a vegetable brush or scrub.

Vegetables and fruits release ethylene gas as they ripen, which helps increase their sugar content and makes them softer. So, if you store potatoes close to other products, they might sprout and soften faster, which might not be what you want, especially if you will not use them all at a go.

Replanting Potatoes The Right Way

If you notice that your stored potatoes have eyes on them, you can decide to replant them. Potato eyes are sprouts that crop up from potatoes when they are ready for replanting.

Potato eyes

Kindly note that all potatoes have eyes. However, some potatoes do not sprout, mainly because of treatment with a bud suppressant chemical such as CIPC carbamate.[10]

If you notice that some of your stored potatoes have sprouted, replant them rather than eat them because they contain toxic chemicals that can make you sick. In addition, potatoes with sprouts do not have their original flavor, which makes them perfect for planting instead of eating.

Before you plant your potatoes, be careful not to break or disturb the sprouts. In addition, if you are dealing with big-sized potatoes, you can cut them into several smaller sections (ensure that each cut section has at least one eye, but it is better to use cut pieces with at least four eyes).

To plant them, place them in an area with adequate sunlight, then cover them with 4 inches of fresh soil (the cut side should face downwards and the eyes upwards). These sprouts will continue growing and emerge from the soil after approximately 14 days and then continue growing until they mature.

---

[10] https://bit.ly/3TURnaB

Once you harvest your potatoes, you might not notice that some may be affected by illnesses and diseases. Also, your potato plants might have some discoloring, or the roots may seem diseased. Read the next chapter to learn about illnesses and conditions affecting potatoes.

Chapter 6: Common Pests and Diseases Affecting Potatoes

As we are about to discover, many diseases affect potato tubers, so ensure you check for any disease symptoms after each harvest season. Doing this will help you identify the diseases in your soil so you can apply effective measures to prevent a recurrence.

Bacterial Diseases

The most common ones are:

Bacterial ring rot

This disease causes the potato leaves and stems to wilt, with the lower leaves wilting first. You will also notice a creamy-yellow to brown rotten ring, which becomes visible when you cut the tuber crossways. The bacteria can enter the tubers through cutting wounds and mainly occurs in warm, wet soils.

To manage the disease:

- Remove all crop debris from the soil before planting and after harvest to prevent soil-borne diseases caused by microorganisms that survive in the soil.

- Keep your tools and equipment thoroughly clean and sanitized, as dirty gardening tools can spread diseases and viruses to a healthy plant from an infected one.

Common scab

This disease causes raised tan, deep brown, or black lesions with a rough texture and tan-colored, translucent tissue inside. The lesions are also circular or irregular in shape. The disease is more common and severe in warm and dry climates.

To control this disease:

- Ensure soil PH levels are below 5.2 because potatoes do best in slightly acidic soils.

- Maintain a high soil moisture content for one month after the stems above the ground swell before the tubers develop. This is critical during the tuber development stages to prevent malformations and translucent tissue disorder (a defect of potato tubers characterized by translucent ends formation).

- Use appropriate fungicides when necessary to treat the seed.

Blackleg

The black leg causes small lesions filled with water at the base of the stems that develop from seed potato pieces. The lesions grow larger to form a lesion that extends from the bottom of the stem to the canopy, and the tissue becomes tender and soaked with water that is either light brown or black. The leaves become wilted with a soft and slimy texture when wet.

The primary cause of this disease is bacteria carried in wounds and on tubers that can easily spread to other healthy tubers when cutting the seed potatoes for planting.

To manage Blackleg:

- Clean your tools and equipment well when cutting your seed potatoes

- Avoid wounding the tubers during harvest to prevent the entry of pathogens through the wounds.

- Try as much as you can to keep the leaves dry when watering to prevent fungal diseases.

Fungal Diseases

Here, your primary concern is:

Potato early blight

Potato early blight causes dark sunken or raised lesions with yellow borders on the leaves or stems. Dark and dry lesions with a leathery texture and yellow border also form on the tubers. The disease primarily affects the crop in wet and dry climates with periods of prolonged leaf wetness and high humidity levels.

To manage early blight, do the following:

- You can apply protective fungicides to your plants.

- Water adequately.

- Store your potatoes in a cool place to prevent the spread of pathogens.

- Plant maincrop varieties are less susceptible to diseases.

Late Blight

This water mold disease affects potato stems and tubers; when left untreated, it spreads fast and results in total crop failure. Potato plants affected by late blight will have large brown or green-gray leaves, and the tubers are shrunken, soft, and discolored.

To manage this mold;

- Harvest and store your potatoes correctly, as discussed earlier in this book.

- Spray fungicides such as chlorothalonil on the soil and plants.

- Grow more blight-resistant cultivars such as;

  o Beyonce

  o Ardeche

  o Jacky

  o Twinner twister

  o Levante

  o Carolus

  o Alouette

Powdery scab

The disease causes white to brown galls (abnormal swellings or growth) on the stolon and roots, shallow swellings covered with brown spores, and raised spots surrounded by the skin on tubers.

To prevent or control it, do the following:

- Avoid planting your seed potatoes in soils that do not drain well.

- Practice crop rotation for 3-10 years.

- Do not plant tomatoes in the same area, as this can lead to a buildup of diseases and nutrient-deficient soil.

Pink rot

The symptoms include a plant that experiences stunted growth leaves that wilt, tubers that tend to rot, dark brown sprouts on tubers, and the tuber turns pink when cut after being exposed to air for 20-30 minutes, which later turns brown and finally black. The disease mainly occurs when potato plants are overwatered late in the season.

To control or prevent this condition:

- Plant your seed potatoes in well-draining soils.

- Avoid wounding your tubers during harvest.

- Avoid overwatering the plants.

Black dot

Symptoms of black dots include small black dots on stems, stolon, and tubers, wilting yellow leaves, and roots that tend to rot under the ground. Poorly drained soils, overwatered soils, and high temperatures are the leading causes of the disease, but the causes are all preventable using means like:

- Crop rotation.

- Adequate watering.

- Protective fungicides should be applied when necessary.

Verticillium Wilt

The cause of this fungal disease is a fungus called verticillium dahlia or verticillium albo-atrum. Leaving this fungus untreated can cause yield losses because it primarily targets the tubers. Besides causing tuber reduction, it also causes stem-end discoloration.

The plant has this disease if you see a tuber with a black or brown ring

To manage this disease, consider doing the following:

- Consider crop rotation. Alternate potatoes with plants susceptible to this disease. These plants include mustard, corn, and cereal crops.

- Ensure you plant certified seed tubers[11] only.

- Select wilt-resistant potato cultivars.

- Regularly remove any weeds.

- Dispose of infected potato plants.

- Avoid over-irrigation.

---

[11] https://faolex.fao.org/docs/pdf/est98479E.pdf

Silver Scurf

A fungus causes this disease. Infected potatoes have superficial silvery patches caused by the separation of the outer layer of skin. These patches darken and enlarge during storage, especially in potatoes stored under damp conditions. Infected seed tubers primarily transmit this disease, but it is also important to note that the fungi survive in the soil for two years.

To manage this disease;

- Rapidly cool infected potatoes. This will delay any disease development.

- Treat infected potato plants with fungicides such as thiabendazole

Other Diseases

Some other potato diseases you need to know are:

Potato late blight

Symptoms of the potato late blight include brown lesions with irregular shapes that spread on leaves and white spores on the underside of the leaf that occur in wet climates.

However, in dry climates, the lesions dry and turn dark brown with dead tissue; later in the infection, the leaves completely rot. The pathogen can last many years in the soil, although the disease emerges when the soils are moist or in cool climates.

To control the disease:

- Destroy the infected tubers to prevent further disease spread.

- Apply appropriate fungicides.

- Avoid watering the leaves.

- Water early in the morning so the plant dries during the day.

- Plant-resistant varieties.

Potato leaf roll

Young leaves roll up and turn pink or yellow, and the plant may experience stunted growth. Aphids transmit the disease, and the infected tubers infect the healthy ones with the virus.

To manage it,

- Try harvesting your crop early in temperate regions so the aphids that migrate later in the season do not affect the plants.

- Apply an appropriate fungicide.

- Remove and destroy any infected tubers and plants so they do not contaminate the healthy plants.

Potato Spindle Tuber Viroid

This potato disease causes stunting and malnutrition. For the tubers, it causes severe cracking. Transmission is easy: it moves from diseased plants to healthy plants or from tools and tractor wheels that touch the diseased plants.

In addition, aphids are another common way this disease spreads (we will discuss more aphids in the next section). This disease is easily noticeable because it affects the tubers and upper leaves.

Although the severity of this disease is high, no biological or chemical means effectively treat this disease. However, you can control this disease by:

- Planting healthy and disease-free seeds.

- Maintain good crop sanitation practices such as disinfection, cleaning, drying, and rinsing.

- Uproot and burn infected plants to avoid spreading the infection to healthy potato plants.

Potato Virus Y

As the name suggests, this virus causes dead spots on potato tubers and leaves. Aphids are the primary cause, which means one infected plant can easily infect other plants.

To manage it;

- Plants potato seeds that have low virus content, such as;

  o Sebago

  o Norwis

  o Monona

  o Kennebec

  o HiLite Russet

  o Belarus

  o Dark red Norland

  o Eva

- Apply mineral oils frequently.

- Sanitize your farm tools every time to reduce the spread.

- Pull out the infected plants.

- Avoid planting the potatoes next to weedy hedgerows and ditches.

- Regularly practice good weed control.

Phytophthora

This disease is a water mold that leads to other severe diseases such as potato blight or late blight. This disease makes the leaves circular, with brown lesions, which are then surrounded by collapsed pale tissue. The leaves also are wet, water-soaked or appear oily. The stems might have dark green-brown or black water-soaked lesions, and the tubers appear sunken and firm, which might extend to some extent into the tuber.

Pests

When it comes to potato pests, the concerning ones are:

Potato and peach aphids

These small, soft-bodied insects are found mainly on the underside of potato leaves, and plant stems. If the aphids attack the crop in large numbers, they may cause the leaves to turn yellow and distorted, as well as cause stunted shoots and spots on the leaves. Also, the sugary substance secreted by the insects can encourage mold to grow on the plants.

To manage your potato crop:

- Prune the infested leaves or shoots if the aphid population is small.

- Check your potato transplants for aphids before replanting them.

- Use reflective mulches like silver-colored plastic to keep the aphids away

- Spray sturdy plants with strong water jets to knock the aphids from the leaves, and spray insecticides if the aphid population is high.

- Introduce predatory insects such as lacewings, syrphid flies, and lady beetles to the plants. These insects are beneficial because they effectively suppress aphid populations and will not harm the plant!

- Spray horticultural mineral oil, pyrethrin, malathion, and azadirachtin. You can also use insecticidal soap on the plant to kill the eggs.

Cutworms

When cutworms infest your garden, you will notice that the stems of young transplants or seed potatoes may appear damaged or severed in the soil. If the infestation occurs when the tubers have grown, you will note some irregular holes in the surface of the tubers.

The larvae that damage potatoes and other vegetables hide in the soil at the base of the potato plants or in plant debris during the day, feeding on the crop at night.

To manage cutworm infestation;

- Remove all plant debris from the soil after harvest or before planting.

- Fit plastic or foil collars around plants to repel off most pests (some might still find their way in eventually).

- Since cutworms feed on weeds mostly, ensure your potato garden is weed-free.

- Place plastic collars around the base of the potato plants

- Sprinkle some food-grade diatomaceous earth around the bases of the potato plants. The diatomaceous earth's texture might seem so soft to the touch, but it cuts through the cutworm's flesh.

- Apply appropriate insecticides and pesticides such as BT—bacillus thuringiensis.

Colorado Potato Beetles

This pest is half an inch long, hard-shelled, has a round, convex shape, and has yellow forewings with ten black stripes that run longitudinally. Its eggs are yellow to bright orange, oval, and laid in clusters of 10 to 30 eggs below the leaves.

To know when Colorado potato beetles infect your potato, look for holes of different sizes on the leaves, specifically around its margins. The pest will have eaten the leaves so much that the skeleton will be the only thing left.

To manage these pests;

- Handpick them.

- Drop its adults and larvae in a container filled with soapy water.

- Crush or remove the yellowish-orange eggs on the underside of your potato leaves.

- Consider crop rotation. Please do this by rotating your potatoes to a new space 200 yards away from it currently. If this is impossible, divide it with barriers such as rivers, roads, and woodlands.

- When you plant your next batch, ensure the seeds are of good quality. In addition, as the potato seeds grow, please practice good nutrition, as discussed earlier in this book, to give them added resilience as they grow.

- Mulch your potato plants with straws. Specifically, strip plant the potato plants in a rye mulch, then mow and push the rye straws over the potato plants as soon as they emerge.

- Plant perimeter trap crops such as sticky nightshade and the African nightshade. These plants attract beetles before they sprout, which means when they grow, the chances of these pests growing will reduce considerably.

- Plant three to five rows of potatoes treated with a systematic insecticide around the garden. This treated border will help eradicate around 80% of the colonizers.

- Use a hand-held flamer or a tractor-mounted flamer on crops under 4 inches; doing this will help kill all or most of the colonizing adult beetles. The flamer will scorch or injure the beetles' antennae and limbs, making it impossible to move along the potato plants.

- Use pesticides such as abamectin, azadiractin, chlorantraniliprole, cyromazine, novaluron, spinetoram, and Spinosad

Wireworms

This worm is a click-beetle larva. The adult is slender, red-brown to black, and 0.25 to 0.5 inches long. The larvae, as seen above, are wirelike, slender, yellow-brown in color, and grow as long as 0.75 inches.

Adult wireworms do not damage the potatoes; the larvae damage seeds and young roots. Therefore, seeing shallow to deep holes in your potatoes could indicate a wireworm infestation.

Since wireworms affect seeds mainly, it is best to manage them before planting. Do not plant if you have uprooted clover, pasture, or grass.

However, if you notice the presence of wireworms in your garden, you can bait them and then remove them by hand (although this works for small populations).

Dig trenches around 4 inches deep and 3 feet apart to lure and kill them. Then, fill these trenches with germinating corn, beans, or peas the cover them with boards.

After a week, check the channels; you will find the worms gathered there, and all you have to do is remove the board, crush the wireworms then replace it.

Flea Beetles

This pest measures around 1.7mm in length and 1mm wide, has black and brown antennae and legs, and mainly attacks the above-ground parts of potato plants and the soil surface because they feed on the lower and upper leaf surfaces.

If flea beetles infest your potatoes, you will see rounded scars measuring around 0.1 to 5 mm in diameter and penetrating the leaves, forming holes. If left untreated, these holes will increase, which will not only affect the potatoes, it will reduce the yield. It may bring about illnesses such as common scabs, fusarium dry rot, Rhizoctonia, and verticillium wilt.

To manage flea beetles:

- Use insecticides such as Seven Insect Killer.

- Use sticky traps to capture the flea beetles, reducing their jumping ability.

Sticky traps

- Spry the leaves with talcum powder to repel them. You can also spray them with a homemade spray made in the following way:

  o Mix two cups of isopropyl alcohol, five cups of water, and one tablespoon of liquid detergent.

- o To apply, spray it on the plant, then cure it overnight.

- In spring, delay transplanting plants by several weeks because this will help cut off their food supply

- Till your garden during fall because this will help unearth any hiding flea beetles and prepare the soil adequately for working during spring.

- Plant hyssop, mint, sage, or/and catnips around the potato garden. These plants help repel flea beetles.

Leafhoppers

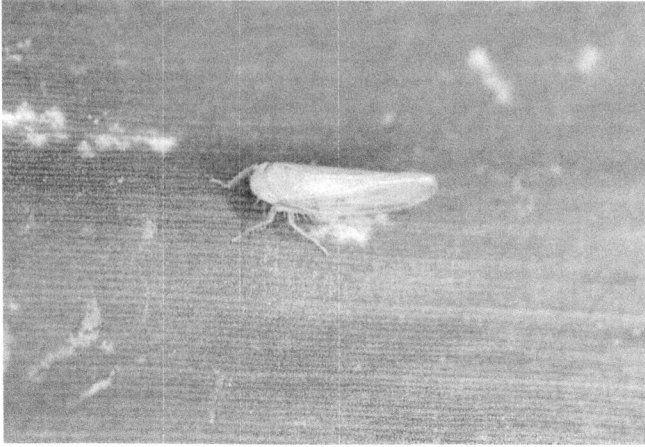

This spindle-shaped, mostly lime-green to yellow-green with some white markings is around an eighth of an inch long. It damages potato plants by sucking the sap, causing an injury called "hopper burns."

Specifically, leaf hoppers feed on the underside of the potato leaves, which causes a yellowing along the margins on the onset and becomes brown at the tips and margins once the damage advances.

To manage leaf hoppers;

- Treat the plants with organic insecticides such as insecticidal soap, azadirachtin, BotaniGard ES, or pyrethrins when pest levels become unmanageable.

- Sprinkle diatomaceous earth on the potato plant to keep the leaf grasshoppers under control. Ensure to cover both lower and upper sections of leaves for efficient results

- Create physical barriers with floating row covers to keep the leaf hoppers away from the garden.

Floating row covers

As we come to the end of this guide, it is essential to know what mistakes you should avoid to end up with a bountiful harvest. Let us look at this next.

Chapter 7: Common Potato Growing Mistakes to Avoid

The most common mistakes you should know and avoid are:

Planting too close

Some farmers pack more seed potatoes into a small space to increase their yield. This results in stunted growth, and the potatoes harvested will be smaller.

To avoid this mistake, measure 12 inches (this works for a space measuring 24 to 26 inches) between one plant and the other. Extra-large potatoes will need the space to increase to 18 inches, and the fingerlings will need around 8 inches.

Over or under-irrigating the potato plants

Some beginner farmers either overwater plants or underwater by ignoring them altogether. You should not under-water your plants because lack of water cause the potato foliage to wilt. On the other hand, they may look droopy if you overwater them.

You can avoid these mistakes by enriching your soil with organic matter or compost before planting your potatoes. Doing this will boost drainage and improve the plant's water-holding capacity. In addition, do not plant potatoes in waterlogged soil or heavy clay.

Not being keen on fertility

Not ensuring the soil is well fertilized may lead to a smaller yield or underdeveloped tubers. To avoid this mistake, mix rotted manure or compost with the soil before planting, then apply about 1½ cups every 10 square feet of garden space. Next, apply diluted fish fertilizer or foliar spray every 3 to 4 weeks (apply every morning or evening).

Conclusion

Growing potatoes may have seemed challenging, but this book has proven that you do not need to be a professional to be good at growing potatoes. Follow the guidelines we have discussed, and you will always be satisfied!

Good luck.

PS: I'd like your feedback. If you are happy with this book, please leave a review on Amazon.

Please leave a review for this book on Amazon by visiting the page below:

https://amzn.to/2VMR5qr

Made in the USA
Las Vegas, NV
09 February 2025